~A BINGO BOOK~

Ecology and The Environment Bingo Book

COMPLETE BINGO GAME IN A BOOK

Written By Rebecca Stark

ISBN 978-0-87386-434-3

Educational Books 'n' Bingo

Printed in the U.S.A.

ECOLOGY & THE ENVIRONMENT BINGO DIRECTIONS

INCLUDED:

List of Terms

Templates for Additional Terms and Clues

2 Clues per Term

30 Unique Bingo Cards

Markers

1. **Either cut apart the book or make copies of ALL the sheets. You might want to make an extra copy of the clue sheets to use for introduction and review. Keep the sheets in an envelope for easy reuse.**

2. Cut apart the call cards with terms and clues.

3. Pass out one bingo card per student. There are enough for a class of 30.

4. Pass out markers. You may cut apart the markers included in this book or use any other small items of your choice.

5. Decide whether or not you will require the entire card to be filled. Requiring the entire card to be filled provides a better review. However, if you have a short time to fill, you may prefer to have them do the just the border or some other format. Tell the class before you begin what is required.

6. There are 50 terms. Read the list before you begin. If there are any terms that have not been covered in class, you may want to read to the students the term and clues before you begin.

7. There is a blank space in the middle of each card. You can instruct the students to use it as a free space or you can write in answers to cover terms not included. Of course, in this case you would create your own clues. (Templates provided.)

8. Shuffle the cards and place them in a pile. Two or three clues are provided for each term. If you plan to play the game with the same group more than once, you might want to choose a different clue for each game. If not, you may choose to use more than one clue.

9. Be sure to keep the cards you have used for the present game in a separate pile. When a student calls, "Bingo," he or she will have to verify that the correct answers are on his or her card AND that the markers were placed in response to the proper questions. Pull out the cards that are on the student's card keeping them in the order they were used in the game. Read each clue as it was given and ask the student to identify the correct answer from his or her card.

10. If the student has the correct answers on the card AND has shown that they were marked in response to the *correct questions,* then that student is the winner and the game is over. If the student does not have the correct answers on the card OR he or she marked the answers in response to *the wrong questions,* then the game continues until there is a proper winner.

11. If you want to play again, reshuffle the cards and begin again.

Have fun!

TERMS

abiotic

acid rain

adaptation(s)

biodegradable

biodiversity

biome(s)

biotic

carbon dioxide

coniferous

conservation

consumer(s)

deciduous

decomposer(s)

deforestation

desert

ecology

ecologist

ecosystem

endangered

EPA

erosion

estuary

extinct

food chain

fossil fuels

global warming

greenhouse

groundwater

habitat

hybrid

landfill(s)

litter

littoral zone

niche

ozone

parasite

photosynthesis

pollution

producers

tropical rainforest(s)

recycling

reduce

renewable

savannas

scavenger(s)

solar panel

tundra

waste

weathering

wetlands

Additional Terms

Choose as many terms as you would like and write them in the squares.
Repeat each as desired. Cut out the squares and randomly
distribute them to the class.
Instruct the students to place the square on the center space of their card.

Clues for Additional Terms

Write two or three clues for each new terms.

_____ 1. 2. 3.	_____ 1. 2. 3.
_____ 1. 2. 3.	_____ 1. 2. 3.
_____ 1. 2. 3.	_____ 1. 2. 3.

abiotic 1. The physical and chemical components of the environment the ___ components. 2. Complete this analogy: biotic : living : : ___ : non-living	**acid rain** 1. Any precipitation that is unusually acidic may be referred to as ___. 2. This precipitation may be harmful to plants and animals.
adaptation(s) 1. It is the process by which an organism becomes better suited to its environment. 2. The shape of a bird's beak or the color of an animal's fur are examples of physical ___.	**biodegradable** 1. If something is capable of being broken down by the action of living things, we say it is ___. 2. ___ products are capable of decaying by the action of living organisms.
biodiversity 1. ___ is the variation of life forms in a specific region, such as an ecosystem or biome. 2. This word is a contracted form of two words: *biological* and *diversity*.	**biome(s)** 1. A ___ is a large area with similar plants, animals and microbes that have adapted to a particular climate or other environmental condition. It is larger than an ecosystem. 2. Grassland, desert, tropical rainforest, savanna & tundra are examples of land ___.
biotic 1. ___ factors are the living components in the environment. 2. The antonym of ___ is *abiotic.*	**carbon dioxide** 1. This gas is absorbed from the air by plants. 2. It is one of the greenhouse gases.
coniferous 1. A ___ forest is composed mainly of cone-bearing, needle-leaved, or scale-leaved evergreen trees. 2. In Russia the ___ forest is called the taiga.	**conservation** 1. ___ is the planned management of a natural resource. 2. Turning off lights when not in use helps in the ___ of electricity.

consumer(s)	deciduous
1. An organism that cannot manufacture its own food from inorganic sources is a ___. 2. A primary ___ is a herbivore that feeds on producers.	1. A ___ forest is composed mainly of trees that lose their foliage seasonally. 2. In a ___ the leaves of most trees change color in the autumn. In the winter they lose their leaves.
decomposer(s)	**deforestation**
1. A ___ is an organism that consumes dead or decaying organisms. 2. ___ break down dead plants and animals. They are important because they help cycle nutrients back into the environment.	1. The permanent destruction of indigenous forests and woodlands is called ___. 2. ___ causes an increase in the concentration of carbon dioxide because there are fewer trees to take in the gas.
desert	**ecology**
1. The ___ biome is characterized by dry conditions and the plants and animals that have adapted to those conditions. 2. Arid land with little precipitation, especially in warm climates, is called a ___.	1. The branch of science that is concerned with the interrelationships of organisms and their environment is called ___. 2. ___ is usually considered a branch of biology.
ecologist	**ecosystem**
1. A scientist who specializes in the interrelationships of organisms and their environment is an ___. 2. A scientist who studies the interdependence of organisms living in the ocean is a marine ___.	1. An ___ includes the living organisms in an area as well as the physical environment. 2. The biotic factors (living organisms) and the abiotic factors (rocks, minerals, soil, etc.) of an ___ function as a unit.
endangered	**EPA**
1. If a species is in danger of becoming extinct, it is said to be ___. 2. The polar bear is listed as a threatened species, which means it is at risk of becoming ___.	1. This acronym stands for Environmental Protection Agency. 2. This agency was established in 1970 to protect the environment against pollution.

erosion	**estuary**
1. The removal of soil, rock debris, and other material from the earth's surface and the transportation of that material is called ___. 2. The action of water, wind, or glacial ice causes ___ of the earth's surface.	1. The body of water formed when freshwater from rivers and streams flows into the ocean and mixes with seawater is called an ___. 2. An ___ is a place of transition from the land to the sea.
extinct	**food chain**
1. If a plant or animal species no longer exists, we say it is extinct. 2. A species become ___ when its last member dies.	1. The eating relationships among species in a particular ecosystem is called a ___. 2. A ___ represents the flow of energy from one organism to another. It usually begins with a plant and ends with a carnivore.
fossil fuels	**global warming**
1. ___ are formed in the earth from plant and animal remains; when burned with air they produce heat or energy. 2. Coal, natural gas and petroleum are ___.	1. It is a sustained increase in the average temperature of Earth's surface. 2. Scientists predict that this will occur because of the increase in the greenhouse effect and that it will cause climatic changes.
greenhouse	**groundwater**
1. The rise in temperature because gases in the atmosphere trap energy and prevent it from escaping back into space is the ___ effect. 2. Carbon dioxide, nitrous oxide, methane & other gases that trap energy, making temperatures on Earth rise are called ___ gases.	1. ___ is contained in the cracks and between the particles of sand, soil and gravel. It is a commonly used for drinking and for irrigation. 2. ___ is stored in aquifers, geologic formations of sand, soil and gravel.
habitat	**hybrid**
1. It is the type of environment in which an organism or an ecological community usually lives. 2. The place where a plant or animal usually lives or grows is called its ___.	1. An automobile that combines a gasoline-powered engine with a rechargeable energy storage system is called a ___. 2. ___ cars are said to achieve better fuel economy than those that use only traditional gasoline-powered engines.

Ecology & the Environment Bingo

© Barbara M. Peller

landfill(s) 1. A ___ is a site for the disposal of waste materials. 2. The techniques of trenching, compacting, and daily covering of the waste with soil are used in sanitary ___.	**litter** 1. Trash that is scattered about is known as ___. 2. It costs millions of dollars to clean up ___ that is thrown from people's cars.
littoral zone 1. The shore line to about 600 feet into the ocean is called the ___. 2. The ___ is divided into vertical layers: the spray zone, which is splashed but not submerged by ocean water; the high tide zone; the middle tide zone; and the low tide zone, which is usually underwater.	**niche** 1. The ecological role of an organism in its community is called its ___. 2. How an organism responds to the resources and competitors in its environment is its ___.
ozone 1. Scientists worry about depletion of the ___ layer because too many UV-B rays will reach the earth's surface. 2. The layer of ___ in the stratosphere filters out a lot of the sun's UV-B rays.	**parasite** 1. A ___ benefits from its relationship with its host, but the host does not benefit and is usually harmed. 2. A flea on a dog is an example of a ___. The flea benefits from drinking the dog's blood, but the dog does not benefit.
photosynthesis 1. The process by which green plants use light energy to turn carbon dioxide and water into carbohydrates and oxygen is called ___ . 2. Green plants use chlorophyll to trap the light energy necessary for ___.	**pollution** 1. The contamination of air, soil, or water with harmful substances is called ___. 2. Factory exhaust and car emissions are sources of air ___.
producers 1. ___ are autotrophs. In other words, they produce their own food from inorganic substances using light or chemical energy. 2. Green plants and algae are called ___ because they create their own food from inorganic substances. Ecology & the Environment Bingo	**tropical rainforest(s)** 1. ___ are found near the equator. They have a greater diversity of plant and animal species than any other biome on Earth. 2. There are four layers in a ___: Emergents, Upper Canopy, Understory, and Forest Floor. © Barbara M. Peller

recycling	**reduce**
1. Taking a product and using all or part of it to make another product is called ___.	1. The 3 R's of the Environment are ___, reuse and recycle.
2. The international symbol for ___ is a set of three arrows moving in a triangle.	2. One way to ___ the amount of waste is to buy products with less packaging.
renewable	**savannas**
1. ___ resources can be replenished with the passage of time. Examples are solar energy and wind energy.	1. Also called tropical grasslands, ___ have only scattered trees and shrubs.
2. Coal, oil, and natural gas—which take millions of years to form naturally—are called non-___ resources.	2. Tropical grasslands, called ___, have long, dry, mild winters and hot, wet summers.
scavenger(s)	**solar panel**
1. An organism that feeds on refuse or carrion is called a ___. (Carrion is the carcass of a dead animal.)	1. It is a devise that collects energy from the sun.
2. Vultures are ___. They feed on the carcasses of dead animals.	2. A ___ utilizes cells that collect sun-light during the daytime and convert the sunlight into electricity.
tundra	**waste**
1. The ___ has no trees, a short growing season, and mostly frozen soil year round.	1. Hazardous ___ comprises toxic chemicals, radioactive materials, and biologic or infectious materials.
2. This treeless plain is the coldest biome. It has a low biotic diversity.	2. Toxic ___ can injure, poison or otherwise harm living things.
weathering	**wetlands**
1. ___ is the decomposition in earth materials caused by exposure to air, water, changing temperatures, and chemical reactions.	1. Marshes, swamps, and bogs are all considered ___.
2. Chemical ___ involves a chemical change in surface rocks. Mechanical ___ involves the breaking of rock into fragments. There is no movement involved.	2. ___ biomes have large amounts of standing water—either salt, fresh, or mixed—for much of the year.

Ecology & the Environment Bingo

Ecology & the Environment Bingo

endangered	adaptation(s)	extinct	tundra	reduce(s)
coniferous	ecosystem	scavenger(s)	groundwater	erosion
producers	littoral zone		estuary	hybrid
solar panel	acid rain	ecologist	wetlands	food chain
fossil fuels	weathering	consumer(s)	abiotic	EPA

Ecology & the Environment Bingo

tundra	recycling	global warming	litter	fossil fuels
food chain	desert	carbon dioxide	acid rain	pollution
ozone	weathering		deciduous	ecologist
groundwater	parasite	littoral zone	waste	erosion
EPA	scavenger(s)	consumer(s)	coniferous	abiotic

Ecology & the Environment Bingo

tundra	ecologist	groundwater	wetlands	producers
weathering	adaptation(s)	biome(s)	ecosystem	habitat
acid rain	scavenger(s)		photosynthesis	biodegradable
littoral zone	ozone	fossil fuels	desert	global warming
abiotic	consumer(s)	coniferous	waste	extinct

Ecology & the Environment Bingo

littoral zone	photosynthesis	extinct	consumer(s)	fossil fuels
greenhouse	desert	ecosystem	litter	producers
estuary	carbon dioxide		reduce	wetlands
ecologist	ecology	scavenger(s)	coniferous	biome(s)
abiotic	EPA	landfill(s)	deforestation	hybrid

Ecology & the Environment Bingo

EPA	reduce	acid rain	carbon dioxide	consumer(s)
greenhouse	ecologist	biome(s)	littoral zone	decomposer(s)
recycling	hybrid		adaptation(s)	extinct
erosion	photosynthesis	endangered	waste	deforestation
groundwater	coniferous	niche	deciduous	estuary

Ecology & the Environment Bingo

biodegradable	photosynthesis	global warming	recycling	hybrid
wetlands	acid rain	deforestation	ecosystem	producers
litter	biome(s)		carbon dioxide	deciduous
coniferous	fossil fuels	waste	landfill(s)	estuary
food chain	ecologist	endangered	niche	extinct

Ecology & the Environment Bingo

endangered	photosynthesis	tropical rainforest(s)	decomposer(s)	groundwater
food chain	extinct	weathering	adaptation(s)	producers
global warming	wetlands		deciduous	biodiversity
littoral zone	desert	greenhouse	tundra	ozone
consumer(s)	coniferous	waste	landfill(s)	biodegradable

Ecology & the Environment Bingo

estuary	photosynthesis	biotic	wetlands	biodiversity
greenhouse	recycling	litter	extinct	reduce
producers	pollution		hybrid	carbon dioxide
abiotic	littoral zone	tundra	deforestation	desert
scavenger(s)	coniferous	landfill(s)	acid rain	food chain

Ecology & the Environment Bingo

deciduous	groundwater	weathering	producers	hybrid
deforestation	recycling	estuary	acid rain	extinct
habitat	endangered		adaptation(s)	biotic
biodiversity	EPA	fossil fuels	decomposer(s)	tropical rainforest(s)
desert	waste	biome(s)	tundra	reduce

Ecology & the Environment Bingo

solar panel	tundra	carbon dioxide	litter	niche
hybrid	biodiversity	ecosystem	adaptation(s)	extinct
photosynthesis	pollution		wetlands	ozone
fossil fuels	erosion	deforestation	waste	habitat
conservation	EPA	global warming	food chain	estuary

Ecology & the Environment Bingo

biodegradable	pollution	acid rain	deforestation	food chain
biotic	habitat	decomposer(s)	deciduous	ecosystem
greenhouse	recycling		global warming	weathering
conservation	producers	waste	coniferous	tundra
biome(s)	consumer(s)	endangered	landfill(s)	groundwater

Ecology & the Environment Bingo

groundwater	desert	habitat	wetlands	deciduous
weathering	scavenger(s)	recycling	landfill(s)	greenhouse
endangered	tropical rainforest(s)		hybrid	litter
consumer(s)	reduce	extinct	tundra	adaptation(s)
pollution	biotic	photosynthesis	biome(s)	biodiversity

Ecology & the Environment Bingo

conservation	reduce	biodegradable	habitat	hybrid
recycling	biotic	photosynthesis	deciduous	ozone
wetlands	carbon dioxide		weathering	tropical rainforest(s)
estuary	waste	biodiversity	pollution	tundra
coniferous	erosion	landfill(s)	endangered	decomposer(s)

Ecology & the Environment Bingo: Card No.13

Ecology & the Environment Bingo

consumer(s)	recycling	acid rain	deciduous	conservation
biodiversity	endangered	habitat	adaptation(s)	ozone
deforestation	wetlands		global warming	carbon dioxide
erosion	waste	photosynthesis	biome(s)	biodegradable
coniferous	litter	pollution	food chain	estuary

Ecology & the Environment Bingo

decomposer(s)	deciduous	acid rain	groundwater	extinct
biodegradable	niche	ecosystem	recycling	deforestation
hybrid	endangered		producers	wetlands
coniferous	habitat	biotic	waste	conservation
food chain	desert	landfill(s)	global warming	weathering

© Barbara M. Peller

Ecology & the Environment Bingo

carbon dioxide	savannas	biotic	niche	parasite
litter	pollution	tropical rainforest(s)	greenhouse	solar panel
conservation	reduce		hybrid	weathering
littoral zone	desert	coniferous	decomposer(s)	tundra
deforestation	habitat	landfill(s)	biodiversity	ozone

© Barbara M. Peller

Ecology & the Environment Bingo

conservation	renewable	ecology	habitat	coniferous
decomposer(s)	deforestation	waste	wetlands	tropical rainforest(s)
deciduous	solar panel		savannas	biotic
EPA	food chain	estuary	acid rain	ozone
fossil fuels	biome(s)	groundwater	tundra	reduce

Ecology & the Environment Bingo

extinct	photosynthesis	biodiversity	deforestation	litter
EPA	conservation	acid rain	hybrid	biome(s)
deciduous	ozone		ecology	niche
pollution	ecosystem	waste	solar panel	global warming
savannas	habitat	fossil fuels	renewable	biodegradable

Ecology & the Environment Bingo

hybrid	biodegradable	habitat	biotic	pollution
decomposer(s)	consumer(s)	niche	groundwater	solar panel
renewable	wetlands		adaptation(s)	acid rain
global warming	savannas	fossil fuels	desert	ecology
producers	parasite	food chain	estuary	landfill(s)

Ecology & the Environment Bingo

pollution	renewable	solar panel	habitat	adaptation(s)
carbon dioxide	weathering	greenhouse	fossil fuels	litter
reduce	tropical rainforest(s)		littoral zone	ecology
EPA	estuary	abiotic	desert	savannas
ecologist	scavenger(s)	parasite	tundra	ecosystem

© Barbara M. Peller

Ecology & the Environment Bingo

biodegradable	EPA	greenhouse	habitat	erosion
reduce	ecology	biodiversity	biotic	endangered
ozone	food chain		renewable	acid rain
fossil fuels	groundwater	savannas	decomposer(s)	estuary
littoral zone	parasite	landfill(s)	conservation	desert

© Barbara M. Peller

Ecology & the Environment Bingo

producers	global warming	ecology	recycling	conservation
litter	solar panel	extinct	biotic	adaptation(s)
biodiversity	wetlands		endangered	tropical rainforest(s)
savannas	EPA	desert	ecosystem	consumer(s)
parasite	biome(s)	renewable	ozone	greenhouse

Ecology & the Environment Bingo

carbon dioxide	renewable	groundwater	recycling	landfill(s)
biodegradable	pollution	food chain	decomposer(s)	ecosystem
global warming	conservation		abiotic	endangered
ozone	scavenger(s)	savannas	biome(s)	desert
erosion	estuary	parasite	fossil fuels	ecology

Ecology & the Environment Bingo

carbon dioxide	pollution	consumer(s)	renewable	biotic
hybrid	landfill(s)	greenhouse	litter	endangered
tropical rainforest(s)	niche		conservation	ozone
erosion	abiotic	savannas	biome(s)	reduce
ecologist	littoral zone	parasite	solar panel	scavenger(s)

Ecology & the Environment Bingo

littoral zone	greenhouse	renewable	acid rain	ecology
ecosystem	erosion	decomposer(s)	carbon dioxide	adaptation(s)
reduce	biotic		abiotic	savannas
niche	EPA	scavenger(s)	parasite	solar panel
landfill(s)	consumer(s)	biodiversity	deforestation	ecologist

Ecology & the Environment Bingo

ecology	renewable	abiotic	litter	niche
global warming	wetlands	biotic	pollution	carbon dioxide
erosion	fossil fuels		solar panel	littoral zone
conservation	recycling	EPA	parasite	savannas
tropical rainforest(s)	deforestation	acid rain	scavenger(s)	ecologist

Ecology & the Environment Bingo

abiotic	biodiversity	renewable	pollution	weathering
erosion	global warming	decomposer(s)	savannas	adaptation(s)
waste	scavenger(s)		parasite	littoral zone
niche	biodegradable	ecologist	greenhouse	ecosystem
conservation	solar panel	ecology	producers	tropical rainforest(s)

Ecology & the Environment Bingo

hybrid	photosynthesis	tundra	renewable	biodiversity
weathering	ecology	abiotic	fossil fuels	solar panel
scavenger(s)	ozone		niche	litter
tropical rainforest(s)	producers	food chain	parasite	savannas
recycling	deciduous	conservation	ecologist	erosion

Ecology & the Environment Bingo: Card No. 28

Ecology & the Environment Bingo

ecology	photosynthesis	niche	decomposer(s)	deciduous
erosion	fossil fuels	greenhouse	tropical rainforest(s)	producers
reduce	renewable		adaptation(s)	abiotic
weathering	EPA	extinct	parasite	savannas
carbon dioxide	biotic	ecologist	biodegradable	scavenger(s)

Ecology & the Environment Bingo

consumer(s)	renewable	litter	deciduous	savannas
ecosystem	niche	global warming	solar panel	adaptation(s)
ecologist	scavenger(s)		tropical rainforest(s)	greenhouse
erosion	biodegradable	ecology	parasite	abiotic
EPA	groundwater	biome(s)	photosynthesis	extinct

www.ingramcontent.com/pod-product-compliance
Lightning Source LLC
Chambersburg PA
CBHW051428200326

41520CB00023B/7394